CLINICAL TEACHING IN CRITICAL CARE SETTINGS
"Challenges & Solutions"

Clinical Teaching in Critical Care Settings

"Challenges & Solutions"

Sameh Elhabashy
CCLN, PhD
Nursing Education Department
Faculty of Nursing
Cairo University

2018

First Printing: 2018

ISBN: 978-0-359-31258-0

Columbia University Press

Columbia University Press
61 West 62 Street
New York, NY 10023
Telephone: (212) 459-0600
USA
www.cup.columbia.edu

Special discounts are available on quantity purchases by corporations, associations, educators, and others. For details, contact the publisher at the above listed address.

U.S. trade bookstores and wholesalers: Please contact;
Sameh Elhabashy
Cairo University
Tel: (+20) 100-7653998; Fax: (+20) 23657190 or email:
Sameh17@med.tohoku.ac.jp

Sameh Elhabashy

To my lovely wife and children

Thank you. Without yan support and persistence, I would have

never accomplished this work.

 # Content

 # Acknowledgements

I would like to express my great thanks and appreciation to my family, parents, colleagues, teachers at Faculty of Nursing-Cairo University who are always willing to provide their support and guidance. I also appreciate the efforts of (Columbia University Press) especially Dr. John P Allegrante for guidance and editing acquisition.

 # Preface

In nursing education, the classroom and clinical environments are linked, because nurses must apply in clinical practice what they have learned in the classroom, on line and through other experience. Clinical teaching is a main part of nursing education. Nurses' exposure to clinical teaching environment is one of the most important factors affecting the teaching-learning process in clinical settings. Excellent clinical teaching is a skill that can be studied, refined, and continuously improved, just as any other procedure. Teaching in the Intensive Care Unit (ICU) comes with unique challenges given the medical complexity of the patients, the time pressure, the diverse levels and professions of learners, and the challenges of communication at the end of life. Identifying challenges of nurses in the clinical teaching environment could improve training and enhance the quality of its planning and promotion of the nurses. Also, This book providing a certain solution, strategies, and plan of management to overcome these challenges. The book utilized different experiences from multiple perspectives in addition to present the latest evidence on (ICU) clinical teaching and incorporates practical tips and examples. I hope that you find this book enjoyable and clinically relevant for all nurses.

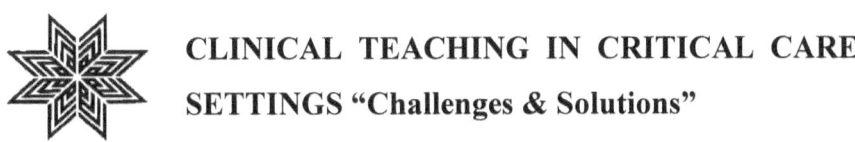

CLINICAL TEACHING IN CRITICAL CARE SETTINGS "Challenges & Solutions"

INTRODUCTION:

Much of the staff nurses make up the health system. So their educations have huge impact on community health. One way to ensure their abilities is providing effective clinical education. So attention to various aspect of clinical education has been done. This book was done to analysis and discuses these dimensions. Teaching in clinical setting presents nurse educators with challenges that are different from those encountered in the classroom. In nursing education, the classroom and clinical environments are linked, because nurses must apply in clinical practice what they have learned in the classroom, on line and through other experience however, clinical setting require different approaches to teaching. The clinical environment is complex and rapidly changing with a variety of new setting and roles in which nurses must be prepared to practice.

Nursing is a professional discipline. A professional is an individual who possesses expert knowledge and skill in a specific domain acquired through formal education in institutions of higher learning and through experience and who uses that knowledge and skill on behalf of society by serving specified clients. Professional disciplines are differentiated from academic disciplines by their practice component. Clinical practice requires critical thinking and problem solving abilities, specialized psychomotor and technological skills and a professional value system. Health care professionals must use critical-thinking skills to solve increasingly complex problems. Educators need to help nurses develop their critical-thinking skills to maintain and enhance their competence.

Nurses' competence is based on the knowledge and skill taught to them. Nursing training is a combination of theoretical and practical learning experiences that enable nurses to acquire the knowledge, skills, and attitudes for providing nursing care. Nursing education is composed of two complementary parts: theoretical training and practical training. A large part of nursing education is carried out in clinical environments. Clinical education forms more than half of the formal educational courses in nursing. Therefore, clinical education is considered to be an essential and integral part of the nursing education program. Since nursing is a performance-based profession, clinical learning environments play an important role in the acquisition of professional abilities and train the nurses to enter the nursing profession and become a registered nurse. Moreover, the clinical area of nursing education is of great importance for nurses in the selection interested area of specialty.

Unlike classroom education, clinical training in nursing occurs in a complex clinical learning environment which is influenced by many factors. This environment provides an opportunity for nurses to learn experimentally and to convert theoretical knowledge to a variety of mental, psychological, and psychomotor skills which are of significance for patient care. Nurses' exposure and preparation to enter the clinical setting are one of the important factors affecting the quality of clinical education. Since an optimal clinical learning environment has a positive impact on the nurses' professional development, a poor learning environment can have adverse effects on their professional development process. The unpredictable nature of the clinical training environment can create some problems for nurses.

 ## Positive Learning Environment in The Clinical Setting

Effective clinical nursing teaching requires good personal characteristics by educators to promote teaching-learning process as well as nursing competence, knowledge, clinical expertise and personality. Accordingly, Clinical teachers should be as role models while supervising nurses and show that they are prepared for teaching. They must also set an example in the clinical fields. It is important to remember that the quality of nurse learning is dependent not only on the type of clinical experience but also on the characteristics and skills of the teacher who facilitates that learning. In order to correlate theory with practice nurses need to gain self-confidence and self-esteem. Hence, to accomplish an effective learning environment, three strategies could be followed. These strategies include knowledge and clinical competence, teaching skills and relationship with nurses.

The researchers' experience in the nursing clinical education reveals that nurses' behaviors and performances change in the clinical setting. This change can negatively affect their learning, progress in patient care, and professional performance. Identifying problems and challenges with which these nurses are faced in the clinical learning environment can help stakeholders solve these problems and contribute to them becoming professional as well as their professional survival. Failure to identify the challenges and problems the nurses are faced with in the clinical learning environment prevents them from effective learning and growth. As a result, the growth and development of their skills will be influenced. Studies show that the nurses' noneffective exposure to the clinical learning environment has increased dropout rates. Some nurses have left the profession as a result of challenges they face in the clinical setting.

Many studies have been done on the clinical environment. Some relevant studies have also been carried out in our country; however, most of them have focused on clinical evaluation or stress factors in the clinical training. One study showed that nurses are vulnerable in the clinical environment and this reduces their satisfaction with the clinical training. Moreover, the nurses' lack of knowledge and skills in the clinical environment can lead to anxiety. Yazdannik and colleagues found that nurses suffered from inferiority complex after entering the clinic. According to a review of the literature, few studies have been done on the challenges nurses are faced with in the clinical learning environment; these challenges are still unknown. Identifying challenges with which nurses are faced in the clinical learning environment in all dimensions could improve training and enhance the quality of its planning and the promotion of the nurses. We aimed to explain the challenges of the nurses in the clinical learning environment.

 **Learning and Teaching in Clinical Settings**

Clinical education plays a large part in a nurse's training. This period is important enough for us to examine its component parts more closely and reach some conclusions that could be useful to new instructors and even to more experienced educators who would be interested in these issues. A more thorough consideration of these issues is essential, since the current nursing program employs a competency approach, and practical training and clinical teaching have large roles to play in nurse competency development. In order to develop a competency, the nurse needs to acquire knowledge, of course, but she also needs to acquire psychomotor, interpersonal, organizational and technical skills; values; a decision-making capacity; and the ability to manage her emotions: in short, she needs to complete a process of personal development that is enriched by experience, and only contact with the reality of practice can provide this experience. Competencies are not only developed by exposure to theoretical knowledge, they are also formed in the heat of the moment, in close contact with situations that resemble what the nurse will encounter in her career.

This is why careful preparation and expert presentation of this training is so important.

We should begin by determining exactly what can be considered clinical teaching, clinical teaching is often confused with practical training. Practical training is experience working in a particular field. The nurse can complete her practical training at the beginning or at the end of her training. The main goal in such training is to give nurses an opportunity to apply the knowledge they have acquired on any given subject in an appropriate practice setting.

The nurse can work independently, but in nursing she is generally under the sustained supervision of an instructor or a "referent" nurse who has the skills needed to coach her and evaluate her performance. Clinical education and the foundations of the profession Clinical education applies the profession's underlying theory to real patient care. Clinical education fosters different types of growth: An excellent opportunity to use emotional excellent opportunity to use emotional intelligence, an ideal place to learn to work independently, an important period in the nurse's progress in terms of the acquisition of skills.

Clinical teaching offers an approach that is tailored to the type and location of the practical training, in addition, the supervising nurse selects and prepares experiences for the nurse, works with her to establishes links between theory and practice, develops the proposed learning program in response to events, monitors the nurse's work, supports her when she has emotional responses, helps her move towards the attainment of program objectives and, generally, assists in her personal development. This process represents a true education, but it is more than that, since the presence of the instructor ensures that patients are receiving quality care. This type of teaching may appear pragmatic and limited to specific areas of care, yet it has the same professional and pedagogical bases as the training received in college. More fundamentally, it even enables the nurse to understand how the basic principles of the nursing profession fit into her day to-day technical work and into how she organizes her work, as shown in the above illustration.

One of the main characteristics of clinical teaching is that it takes place in a working experience, where the nurse is constantly called upon to think about her experiences and adapt her activities in response. The professor's involvement is therefore much greater than simply fulfilling a role monitoring the practical training; it comprises training, facilitating and supervising. The role falls under a double mandate: she is an instructor who is responsible for the nurse as well as the nurse who is responsible for the client being cared for by the nurse. Because of these requirements, the role is a complex task split between concerns for the nurse's learning process and a proper and ethical delivery of patient care. At this point it is worth noting some basic principles, including the need to prepare for this practical training (or as we sometimes call it, this "laboratory-hospital") well ahead of time. This implies that the nurse must already have acquired the knowledge she will be applying in the clinical setting in her work with patients. It goes without saying that one cannot apply what one does not know. This preparation simply represents an exercise in professional responsibility on the part of both the instructor and the nurse, since the clinical setting is where the nurse learns to use the theoretical and practical knowledge acquired in the classroom and the laboratory and apply it under new conditions. (See also learning in a Laboratory College: An Educational Practice That Deserves a Higher Profile.

The Clinical instructor's preparation includes: clearly defining learning objectives and communicating them to nurses, a prior, judicious selection of learning experiences: that must not overburden her with stress, which would only interfere with their learning, that represent good opportunities to learn, prior identification of the patients' health problems and of the main interventions required in patient care; with this preparation behind her, the instructor can provide better supervision.

Another important issue is that once this knowledge has been mastered at college, it should not be necessary to continue the learning in a clinical setting. When the nurse commences her practical training, she has already advanced in her professional development, reaching the point where she needs to apply what she knows. Practical training has a special role to play in her learning. The instructor can nevertheless save learning experiences for the clinical setting that would be impossible to provide in a classroom setting, reserving this special time for the nurse to make observations, undertake interaction exercises or add additional information. In summary, none of the experiences that a nurse can have in college (whether in the classroom or in the laboratory) should be included in clinical teaching. Theoretical or technical learning and exercises that can be completed elsewhere should not be included, as they are not a good investment of this precious clinical time.

 ## The Preferred Instructional Approach

Clinical teaching, like all other teaching, must be based on a clearly identified instructional approach founded in principles that are adapted to this particular reality. Work in a real practice setting requires active teaching aimed at building the nurse's knowledge base and based on principles of group practice. Since nurses work in teams, training that prepares them for this type of action is particularly appropriate. A socio-constructivist pedagogical approach has been proven to meet the requirements of practical training. The group plays a very important role in this approach, since the group values the construction of knowledge by both the individuals in the group and the group itself. The strategies associated with maintaining an active group dynamic are necessarily dynamic and attractive.

A learning process in which the nurse builds her own internal representation of knowledge, a personal interpretation of her experience. This representation is subject to constant change; its structure and links form the basis of the future knowledge that will eventually be added to it.

Constructivist teaching: defined as training and clinical learning require serious preparation by both the instructor and the nurse. For the instructor, this means providing the nurse with adequate knowledge about the scientific, technical and interpersonal context so that she can meet the demands of the clinical experience. In this kind of program, preparing a nurse does not only involve communicating scientific and technical concepts; at some fundamental level it also involves developing her self-actualization, ethics and professional commitment. Even though many instructors may have a very good understanding of how to prepare a nurse's practical training, we will now briefly summarize the main elements of this important work.

Role of Clinical Nurse Instructor (teacher).

The clinical instructor evaluates the nurse abilities and reinforces learning and the performance skills. They also can play an important part in the developing and enhancing nurse's self-confidence and learning outcomes. According to Reilly and Obermann in 1992 clinical teaching is a form of interpersonal communication between two people a teacher and a learner. Hence a clinical teacher may reduce the nurse's anxiety and promote the nurse's ability to apply energy creatively and to achieve learning goals.

Clinical teachers were described as caring mothers in their caring roles in guiding, supporting, informing, translating, sustaining, negotiating, reinforcing, transforming and releasing nurses through their clinical practice. The role of the clinical teacher is very important in facilitating nurses' learning, especially in the clinical settings where uncertainty abounds. The "caring mother" role of the clinical teacher enable nurses to become actively responsible for making learning a formative, stimulating experience.

The roles of clinical teaching members could be summarized as to be skilled, experienced nurse to maintain and improve standards of patient care. The clinical instructor should be concerned to help learners develop their potential as nurses. This could be achieved through building good relationships, counseling supporting and advising. They must demonstrate expertise in caring for patients because patient's life or certainly his well-being could be at risk according to. They also must show skills in teaching create a positive climate for learning and be alert to the learning opportunities in the ward.

Instructor's Preparation for the Practical Training:

• Clearly define the training's objectives.

• Develop instruments for observing the nurse's performance.

• Carefully select appropriate experiences for the nurse: not too difficult or stressful.

• Select a setting for the training and situations that represent good learning opportunities: organization, role models, etc.

• Identify the client beforehand, including the inherent problems and the kind of care that the client will require.

• Request client authorization.

• Prepare clinics and educational strategies.

• Identify complementary work for the nurse to complete.

Instructor's preparations Clinical learning is a serious undertaking, since this is a testing ground where the nurse comes into contact with the reality of care, with all its attendant rituals and demands. It is also where the nurse confronts a real practice setting, with real human beings. For ethical reasons, patients are not treated as guinea pigs, so nurses must have already attained a threshold level of performance in Laboratory College. This is the justification for meticulous preparation, on the part of both the instructor and the nurse, for this critical experience. More specifically, these preparations include:

- Communicating a clear definition of the clinical learning objectives to nurses, setting a precise level of performance needed in their next practical training. When we know where we are going, we have a good chance of getting there!

- Developing observation and evaluation instruments: a template that the nurse can use to gather information from the patient, supplementary exercises to develop her sense of observation, a document for recording her clinical experiences, a book covering her practical training and a logbook. A portfolio may replace some of this work. The instructor must also have an instrument for observing and evaluating the nurse.

- Judiciously selecting clinical experiences of increasing difficulty that are the best experiences possible for meeting the objectives of a practical training period. These experiences must be appropriate to the nurse's emotional development, problem-solving skills and ability to exercise care activities. A situation that is too complex or emotionally too difficult will cause her a significant amount of stress, consume all her energy and overwhelm her capacity to learn.

- Studies has stated that beyond a certain level of tension, we enter into a zone of declining performance; i.e., beyond a certain threshold, we begin to become less effective. This is what happens to a nurse who faces an experience that is too technically or organizationally demanding and, above all, too emotionally upsetting. The nurse can no longer be certain of succeeding and the experience does not confirm her personal abilities, an essential Patient information to be collected beforehand by the instructor from the patient patient's file: Medical diagnosis, surgical intervention Appliances: stoma, prosthesis, orthosis, glasses, etc. Medication: dosage and schedule Treatments: with schedules and details Culture: language spoken, characteristics Lifestyle: for personal care, elimination, meals, wake-up, bedtime and sleep Position: partially sitting, lateral, etc. Exercises: breathing, movement, etc. State: pain, fatigue, discomfort, means for relief characteristics: preparation for the O.R., for transfer, etc.

Inappropriate learning experiences lead to a sense of not being up to the task, and this leaves the nurse feeling discouraged. Selecting situations that offer good learning potential.

- What good will practical training serve if it offers nothing to be learned?
- What will a clinical setting have to offer if the caregivers cannot be taken as valid role models?

Practical training in such a unit would represent time lost or even a bad influence on the nurse. It is also worth remembering that the nurses are first and foremost there to learn, and that their primary role is not as "providers" of care or extra workers who assist the existing staff. The choice of a setting or department for the practical training is therefore one of the most important issues in its success. Before having a nurse work with a patient, the instructor must learn about the patient's problems and identify the main interventions he or she will require. This involves visiting the hospital department and conducting an in-depth review of the patient's file in order to learn about his or her health problem, treatments, medication and examinations, reactions (such as appetite), quality of sleep, the amount of pain experienced, mood, habits, etc. Last but not least, the instructor needs to meet with the patient and ask his or her permission, explain the level of personal care and performance that can be expected, and the monitoring and coaching that will be in place. The patient may well refuse the request, but this rarely occurs once the advantages have been explained. In summary, before the practical training begins the instructor must have a good general idea of what the nurse will be doing, the main problems she will confront, and the teaching and explanations that will be needed in order to respond to different aspects of the situation.

The instructor will also have to plan the various educational strategies she will be using. For example, to ensure that the nurse knows the organization and its staff she can ask new nurses to make a quick sketch of the department's organizational chart. A quick and informal plan of the care unit may also help the nurse find her way around. Nurse's preparations the nurse also needs to prepare for her clinical learning if she wants to get the most out of the experience and be sure that she is effective in the workplace. This preparation is also important for the safety of patients. Ideally, she must know ahead of time (i.e. the night before) what type of patient she will be caring for. This will allow her to review the health problem, the related surgery (if surgery is required), the types of tests required or the current treatment. But above all, she must use the information to think about the plan of care she will need to develop for the patient. This will not only make her work easier, it will surely make it better. Under this approach, each situation the nurse faces will involve new learning. Some research and reflection will also help her absorb this new clinical and technical knowledge into her mental structures and transfer them to other situations as required.

 ## Orientation for the Clinical Setting and Experience

In order to be comfortable and effective in the clinical setting, the nurse must first receive a good introduction to the unit of care where she will be working. This will help reduce her stress through a more gradual adaptation to the general climate in the department, the staff and the physical surroundings. This orientation is intended to make her comfortable with the unit, provide basic knowledge about how it works and even give her an idea of the equipment used and where to find the storage areas to making the experience of her first visit all that more concrete. One approach is to ask her to complete a short exercise to reinforce the 7 information she received during the introduction. For example, she could be asked to show, in a kind of a treasure hunt, where she would find various items that she will need in her work or which members of the staff she will need to see for certain needs. She can be given an organization chart (if she was not asked to prepare one herself) with the names of the people in each position in her unit. She can also be asked to observe certain aspects of how the unit operates. It is useful to set aside some time after this first formal contact with a hospital, time for the nurses to get together and discuss the experience, share their fears and talk about their emotional responses.

Assigning Clients to A Nurse

Assigning clients to a nurse for her practical training is an important step in the preparation of this learning experience. The instructor can decide to select the clients and assign them to the nurse herself, based on her judgment of the nurse's abilities, the challenges she needs to face. It is also possible to let the nurse choose the clinical experience that best responds to her learning needs, since she must cover a certain amount of ground in the course of her education. What is important is that the nurse has access to all the knowledge needed by nurses. Also, in this case, practical training experiences must be selected from among a certain number of clients selected by the instructor.

Clinical Education

Clients may be assigned to nurses in several ways:
• Preceptorship: A more formal pairing with a member of the staff. An expert oversees the work of a new professional in order to help her learn.
• Membership on a care team: One or two nurses are assigned to a team and participate in its usual operations.

The goals are to have the nurses' model behavior, develop technical and organizational skills and learn to work on a team. To prepare a nurse for her work with clients and according to her level of training and her ability to work independently, the professor may either provide her with the information needed for care (as described above) or suggest that the nurse does the research herself (even if this means completing the information that she has collected at some later time). The attribution of patients to nurses can take several forms: individual, paired, alternative or preceptorial assignments. To this list must also be added integrating the nurse into a care team. Individual assignment and assignment in pairs Individual assignment is probably the most common method. It consists of confiding the care of one or more patients to a single nurse.

Assignment in pairs involves confiding the care of one or more patients to two nurses who work together. This approach alleviates anxiety and helps nurses develop an approach based on sharing and mutual support. It can be particularly useful as a first assignment. In addition, when faced with a complex situation that tests her abilities, the nurse can be paired one-on-one with a staff member, with a more advanced nurse or with a nurse who has demonstrated superior performance. The instructor thereby offers assistance and a role model, which, based on the writings of Albert Bandura, we know is a very effective learning experience. His theory of vicarious learning applies to what a nurse learns in parallel to the material that the instructor explicitly provides, by observing the instructor or another nurse at work and learning by modeling their behavior.

Client attribution Clients can be assigned to the nurse in several ways:

• Individual assignments: one nurse for one or more clients

• Paired assignments: two nurses for one client (a difficult case) or for several clients; nurses can also be paired with a staff member or with a more advanced nurse

• Alternative assignments: one nurse takes a specific role with her classmates (help with research or with care) when opportunities for clinical experience are limited. A pared assignment can also prove useful when resources are limited. There are several ways it can work. For example, two nurses can care for three clients. This type of assignment is particularly useful at the start of a course, as a way of easing nurses into their new roles.

Alternative assignment This form of assignment can be used when a nurse has different learning needs or needs more complete learning experiences, or when the learning experiences offered on a particular care unit are limited or do not offer enough variety. The nurse can be asked to assist some fellow nurses in care delivery. This could involve research on medications and the types of tests and treatments required, information that she must then provide to the interested parties. The instructor can also assign the role of preparing and carrying out activities for paediatric or psychiatric clients. In some cases, it is also possible to ask a nurse to prepare, administer and record certain types of medications, if this requires a specific type of supervision at this point in the nurse's training.

 **PRECEPTORSHIP**

Preceptorship refers to a more formal paired assignment with a designated member of the care unit. This is not a new approach; it used to be known the "sponsor" system. A preceptorship pairs a "novice" with an "expert." This makes it easier to develop the nurse's clinical competencies and dampens the shock of dealing with real-world situations. The system can be made more effective if the experience is well planned and the selection of the preceptor is apt. The preceptor must be interested in taking the nurse's learning objectives into account and, in some ways, take responsibility for them. Our hospital systems are more and more open to this type of arrangement, which is already used for the internship period during vacations. Role played by the mentor Teaching, professional and technical support of nurses Identification of helpful information and resources Supervision of how well learning requirements fit the results Assistance in contacts with other stakeholders Openness and availability: lending a ready ear Psychological support when the nurse runs into problems Evaluations of the mentoree's progress, suggesting ways to correct or improve performance Adding a nurse to a care team Another way to give a nurse contact with patients is to make her a part of a care team. This allows her to begin her nursing experience with "caregivers" in an entirely real-world context.

Preceptorship provides:

• A thorough learning experience, progressive and well founded in reality.
 • A progressive shouldering of responsibility.
• Tailored nurse support.
• Constant supervision.
 • Optimal protection of the public.

Role of Clinical Nurse Preceptor.

The clinical preceptor has a crucial and essential role in developing the clinical knowledge and skills of the nurses, as a large part of their education involves clinical practice. Clinical preceptors have a dual function care of patients and care of learners. Moreover, growth in applying the theoretical knowledge and skills in the clinical setting is a major task for clinical nurse teacher. In addition, the clinical instructor plays an important role in providing appropriate information which is suitable for the clinical part. Also, he or she gives feedback to the nurse and provides follow up support.

These learning experiences occur as much by modelling the behavior of other nurses as through an immersion in the practice setting. Organizational components and knowledge of psychomotor and interpersonal skills are experienced as interrelated phenomena, closely connected and continuous, a supplement to her learning. The nurse is able to see how a nurse organizes her day, sets priorities and resolves problems as they arise, information that the nurse then applies to her own activities. This form of pairing is effective, since it allows the nurse to learn in a real practice setting, but it also has the advantage of providing her with the emotional support she needs to reduce the stress of starting out in a clinical setting. It is nevertheless very unfortunate that not all practice settings are open to receiving several nurses in this manner.

We all know that nurses are overwhelmed with work, and that having a nurse on the team may, at times, make situations more difficult. It must nevertheless be pointed out that a nurse with little experience is still able to provide considerable assistance on a care unit. However, if a nurse is given a placement on a team, the team's head nurse needs to have a very good understanding of the instructor's objectives. Without actually doing the instructor's work, the head nurse can play a support role. 10 High points in clinical learning In the course of a practical training experience, nurses learn mostly in their contacts with clients, but they also learn during clinics held by the instructor. The pre-clinic, as all instructors know, is when professors provide detailed information or explanations about the client or clients selected for the nurse. It is also an opportunity to direct the nurse to sources where she can look for more information in order to better understand different aspects of the patient's condition.

Sources of Pleasure at Work for ICU Nurses

According to Alderson's approach, pleasure at work refers to the state of psychological well-being that workers experience when their work satisfies their desires for recognition and thus allows them to build their identity" this approach considers pleasure at work—like suffering—in collective terms. This section sets out the main sources of pleasure at work of ICU nurses mentioned in the literature. It should be noted that the sources of pleasure related to the organization of work are nowadays in continual decline for ICU nurses. In the literature they are usually expressed as factors that give rise to satisfaction and are very often the opposite of the sources of suffering.

A number of authors maintain that ICU nurses love challenges and like taking risks; in other words, they see stressful situations as motivating, as challenges rather than threats. Lazarus defines a challenge as a situation that gives a person a sense of wellbeing while producing stress or tension. In fact, Carpentier-Roy notes that the constant challenge of danger and of the technology is one of the most common justifications nurses give for choosing intensive care; they enjoy the risk and the unpredictable nature of the work, where death is constantly lurking.

 # HIGH POINTS IN CLINICAL LEARNING

High points in clinical learning can occur: At the morning orientation, at the clinical meeting during the day, at the post-clinic at the end of the day, at times of interaction with the nurse throughout the day when the nurse is asked questions, given explanations and receives specific comments, or when her performance is monitored or assessed. The clinical meeting takes place during the day at a time that is best for the care organization. It can take place while patients are resting or when there are fewer demands on the staff's time. This clinic is used to:

- Discuss emotions aroused in the nurse and her impressions of the care experience.
- Provide her with support and understanding.
- Learn about any problems encountered in care: the patient's mood, a technical or organizational problem, specific concerns, etc.; 18 Strategies for practical training Strategies must be adapted to the nurse's level of academic achievement.
They can be used during clinical meetings or as a complement to her clinical work.

During the clinical meeting: Exercises that involve observing general activities or the client An oral care process: have the nurse describe the stages in an oral care process for a given client A report on an exchange with her client A clinical reasoning workshop Complementary exercises: Training, report, Portfolio Diagrams, organizational chart, unit plan, Concept map or tree:

- Provide additional information.
- Direct the nurse to useful resources.
- Motivate nurses to do their research and learn.
- Give them direction in their work.

The post clinic provides the instructor with an opportunity to follow up on both cognitive and affective aspects of learning. This is where the instructor encourages the nurse think about her experiences and establish links, reflect, make choices, generalize about the experience and organize her knowledge about the client and the clinical setting. In addition, it is where nurses share experiences with each other. The post clinic gives them an opportunity to talk about the day's experiences and receive any support or explanations they might require. There may be times when nurses have suffered a loss that has had a profound effect on them and that requires emotional support from the instructor. But there may also be conflicts with patients, with staff 11 members or even with other nurses, at which time the instructor will need to encourage dialogue and reconciliation. Clinical lessons given during the course of the day. In reality, the instructor teaches throughout the day. The instructor helps nurses attain their learning objectives through questions and explanations that are given on the fly. Since she is often present when nurses are providing patient care, she can evaluate their performance or suggest improvements at that time, but it is important to remember that this teaching has its basis in the questions she asks rather than in any answers provided.

Learning strategies in these different areas just as classroom and laboratory learning require appropriate teaching strategies, strategies must also be developed to help the nurse learn in a clinical setting. The instructor must find the means to develop the nurse's sense of observation and encourage her to conduct research and develop an ability to think critically. She will use these skills to continually establish links between her considerations for the importance of her actions and the practical experience itself. In practical training, learning is fostered by developing this unending thought/ action feedback loop.

Nursing diagnosis concept tree (deduction) what data need to be collected? Alteration of the mucous membrane of the mouth Symptoms To this end, the professor may ask the nurse to complete an observation exercise that consists of focusing on one aspect of the client, such as facial expression and non-verbal behavior, what the client says, his or her reaction to pain, etc. This is essentially a small data-gathering exercise performed with the nurse. The instructor could also ask for a summary of the client's status and treatment (pathology, surgery, investigative tools, treatments, medication, etc.): i.e. a short study of a real case or a concept map of the health problem. Another approach is to have the nurse conduct research on investigative tools or on the medication prescribed for her client. Using inductive reasoning, she can determine the underlying principles of some of the care required by her client. This is another way to encourage the nurse to adopt more complex reasoning processes; it allows her to bring these principles to a conscious level so that she can see how, for example, the principles underlying respect, comfort, communication, asepsis, safety, and saving energy apply to client care.

Nursing process and diagnosis the nursing process is still the best way to acquire this type of learning, even though we may regret the fact that too often the process is completed only after care has been provided. (However, it must be acknowledged that a nurse cannot prepare her complete process at the start of the day.) We can nevertheless ask the nurse to run through the process in her mind or verbally, at least to gain a general sense of it; to summarize her data gathering; to make one or more tentative diagnoses; and even to suggest any interventions that she believes are merited. This exercise sets objectives at the start of the day, even if corrections will be required along the way once she has accumulated more data. It should be understood that in a clinical setting the nursing process actually begins in the morning as soon as a client is assigned to the nurse, and continues throughout her work day. The process begins with care; only her final description of the process and her evaluation should take place later.

The nursing process is like an itinerary that the caregiver has to follow. The instructor can also use some "drill" exercises to assist learning of, for example, the nursing diagnosis, and employ certain inductive or deductive reasoning strategies to identify the questions required in a specific situation, to state the objectives for a specific diagnosis, to determine the nursing diagnoses themselves, and to suggest certain interventions that can be developed on the basis of these nursing diagnoses. For better learning, we should above all emphasize clinical decision-making, since clinical judgment and the decision-making process can be taught and improved through educational practices. One approach is to ask the nurse to prepare a list of the problems she has noticed in the client or, for one of these problems, list all the types of care that could be provided. The drill or intensive drill or intensive exercise. All that we need to add is the appropriate opportunities and use appropriate pedagogical strategies.

 **Consequences of the Current ICU Work Environment**

A 2001 study by the Ordre des Infirmières et Infirmiers du Québec reports that the unmet need for nurses has grown in various sectors and has affected the quality of nursing. Indeed, in a descriptive study, Morrison et al. reveal that errors tend to increase in ICUs in which very acute clinical situations are combined with a shortage of nurses and inadequate supervision. Beckman and Gillies note that the failure of early detection of deterioration in patients' condition (failure to rescue) in ICUs tends to be exacerbated when there are not enough experienced nurses. Some authors point out that staffing levels and the collective competence of nurses may explain many aspects of patient morbidity and mortality. It is interesting to note in this regard that the difficult working environment engenders individualism, hampers solidarity and undermines or limits the formation of true working groups. Such situations produce feelings of impotence in workers, leading them to turn in on themselves and disengage from their work.

Consequences for nurses' health

The consequences of the difficult working conditions in ICUs on the health of nurses must not be ignored. Unlike studies that seek mainly to identify or describe the methods ICU nurses use to adapt to stressful situations at work, the approach seeks to understand the relationship between the way work is organized and the mental health of the workers. Vézina points out that from the standpoint of this approach, equilibrium on three dimensions— fulfilment through work, identity building and recognition by others—helps workers maintain their mental health. The author points out that by letting workers use their creativity, by encouraging initiative and by allowing them some autonomy, work enables people to fulfil themselves and develop their identity. In contrast, an organization of work that does little to foster creativity or initiative and leaves little room for negotiation, hinders identity building and consequently undermines psychological wellbeing. When workers no longer have the opportunity to reflect on their work or to discuss situations they encounter with each other, the result is often that the work loses its meaning. The present situation in ICUs seems to be of this second type.

Some studies on nursing have more particularly investigated the relationship between the environment of ICUs and the health of the nurses practising in them. Thus, as early as 1979, Oskin reported that in the United States measured stress levels of ICU nurses were associated with a health risk. More recently, other authors have pointed out that the high level of stress that ICU nurses experience engenders signs and symptoms of both a physical (e.g., headache, skin rash, gastrointestinal disorder, weight fluctuation) and psychological (e.g., loss of morale, burnout) nature that are associated with a loss of productivity.

Cronqvist et al. maintain that when the workload is heavy, ICU nurses experience more stress, and absences for personal illness increase. Furthermore, Tummers et al. and Le Blanc et al. note that the heavy workload and prolonged stress of ICU nurses is correlated with burnout. Bakker et al. observe that burnout-which Robinson and Lewis define as related to continual exposure to stressors at work and to the use of ill-adapted coping strategies- is a serious problem for ICU nurses. In their study of burnout contagion, Bakker et al. report that ICU nurses are more likely to present the problem if they work in a unit where the condition is highly prevalent. In addition, some authors point out that moral distress is a common response in ICU nurses who face ethical challenges.

Often, ICU nurses' time and energy is taken up with tasks beneath their skill level. Moreover, in a study that does not deal exclusively with ICUs, Bourbonnais et al. found that nurses lack the time to practice aspects of their role that have a positive impact on the quality of care. The result is work comprised almost exclusively of technical actions that leave by the wayside the relational aspects of care, which, on a day to- day basis, give meaning to a nurse's work. Other authors add that this type of organization, which results in major work overload and a range of frustrations, helps increase turnover and burnout rates in ICUs.

Consequences for Work Relations

Current circumstances in ICUs have produced major changes in nurses' interpersonal relations by undermining trust between colleagues and vis-à-vis management. As the staffing situation continues to deteriorate, nurses, who have little say in decision making and are subject to mandatory overtime, may suspect that management is not making sufficient effort to improve their lot. Indeed, according to Mishra and Morrisey trust is closely related to employee participation in the decision-making process and to the sharing of information. Employees who are little involved in decision making may suspect that information is being kept from them, and so feel a growing distrust of management. Dejours maintains that the lack or loss of trust by workers has an impact on team collaboration, mobilization and recognition.

Vézina points out that the weakening of social relations between people leaves them bereft and disoriented, making peer recognition all the harder. Vézina notes moreover that work situations that combine greater effort with little reward elicit both

31

emotional and physiological pathological reactions. This combination seems to characterize the situation faced by a number of nurses working in ICU.

To sum up, the studies we reviewed suggest, from several different standpoints, that the three dimensions posited by Vézina fulfilment at work, identity building and recognition by others- are not in equilibrium in ICU nurses. In fact, the sources of suffering at work analyzed in the literature may be considered damaging for ICU nurses insofar as they considerably reduce any pleasure the nurses derive from their work and cause it to lose meaning for them.

The difficult environment and rigid organization of work intensify the factors for dissatisfaction in ICU and create an imbalance between the nurses' sources of pleasure and of suffering. According to Carpentier-Roy, the rigidity of the organization of work limits the nurses' potential to find emotional release, resulting in accumulated energy that induces "a feeling of displeasure and of suffering that can be accompanied by various psychological disorders." Since their capacity for regulating suffering and defenses is diminished by this rigidity, many ICU nurses feel threatened and implement defensive strategies that lead many of them to be frequently absent and even to leave the ICU in order to preserve their mental health. In fact, Vézina and Malenfant maintain that sometimes individuals have no choice but to leave a job in order to reduce the tensions created by their work. The departure of many nurses destabilizes teams and weakens any working groups that may exist. The repercussions for the working climate are negative; they damage the image of ICUs and make work in such units less attractive to nurses. The effect on nurse recruitment and retention is substantial, and the suffering of the nurses who remain on the job is exacerbated. Some of them end up leaving these difficult working conditions. This situation aggravates the initial problem and reinforces the suffering that further saps their quality of life at work.

It is important and urgent to break the vicious cycle described above, otherwise the accessibility of intensive care will be seriously affected, and the sustainability of the system threatened. However, the analysis does point to a number of avenues for thought and action roles, which will be presented below.

 ## Knowledge and Clinical Competence

Clinical teacher must have mastery knowledge of the subject matter. He/she should have surrounded broad knowledge which links between the various theoretical knowledge nurses have learned in the classroom and the practical milieu. For example, say the nurse has just collected a urine sample from a patient, He/she could be asked to explain the anatomy and physiology of the kidneys and common microbes found in urine. This transfer of learning and connection could be drawn from their Human Biology and Microbiology lectures. Through this nurses will capture what it would have been like for them to conduct an intimate procedure.

Though, the clinical teacher's theoretical and clinical knowledge if used in the practice of nursing and attitude toward the nursing profession will influence the teaching effectiveness.

Like knowledge, the effective clinical teaching requires competence in clinical nursing practice as well. The maintenance of clinical competence is essential in assisting nurses in development of knowledge and skills and providing expert supervision in clinical setting. Clinical is the only setting in which the skills of history taking, physical examination, clinical reasoning, decision making, empathy, and professionalism can be taught and learnt as an integrated whole. Therefore, an effective teacher should present information in an organized manner, gives clear explanations and directions to nurses, answers questions clearly, and demonstrates procedures and other care practices effectively. Good teachers also, clarify ideas, emphasize important points during teaching and motivate nurses through active participation throughout their teaching practices. Furthermore, in clinical setting teachers may need to assimilate knowledge starting from the specific towards the general. For example, when teaching about diabetic care, the teacher needs to focus on a specific care required by a diabetic patient such as how to actually administer the insulin injection rather than teaching about the general care of a diabetic patient.

In addition, clinical teachers need to identify individual nurse needs and learning styles and plan supervision accordingly. An example was given by Henderson in 1995, nurses with a predominately extroverted personality may in their zest to please the teacher state that they understand something being explained when in actual fact they do not understand. Teachers who do not pick up on this may fail to facilitate meaningful learning. One strategy that Henderson has mentioned with extroverted nurses is to get them to present to their peers what they have understood from the teacher's explanation. Similarly, introverted nurses may appear to be disinterested and being uninvolved because of their quiet and introspective manner. This is not usually the case and all the teacher needs to do is to gain their trust and give them more time to 'open up' and provide opportunities for them to share their knowledge with others adds Henderson in 1995.

Relationship with Nurses

Facilitating learning involves interacting between teacher and nurses and the ability of the teacher to interact with nurses is a very critical teacher's behavior. Many studies by explore the personal factors affecting and the relationship between these two aspects. And concluded that the most important factor with regard to the teaching effectiveness of clinical nursing teachers was "Harmonious interpersonal relationships".

Clinical preceptor could enhance nurse learning by having an unconditional positive regard for them. Teachers also need to be sensitive to nurse's feelings and problems and convey confidence in the nurse's ability to learn. Henderson in 1995 has given a good example of mutual respect of nurses in situations where teachers need to correct nurses. They need to clearly convey to nurses that it is their behaviour that the teacher is not happy with, rather than being unhappy with them as people. Nurses should not be made to feel that their personhood has been put on the line which could damage their self-esteem as highlighted by this comment.

Implications for Nursing Practice

The issues and trends affecting nursing care today are increasingly complex and dynamic. It is widely believed that nursing leads the whole quality movement in health care. Nurses provide the majority of patient care therefore; they must be empowered with good preparation to improve care and service in order to maintain quality of patient care. Graduating highly qualified nurses will positively and effectively reflect on all rendered services. Based on that, the need to improve clinical nursing education should become a major concern.

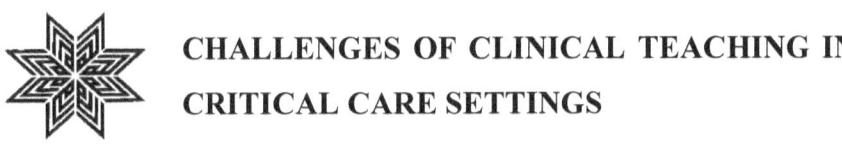

CHALLENGES OF CLINICAL TEACHING IN CRITICAL CARE SETTINGS

The challenges of clinical teaching nursing skills and lifelong learning from the standpoint of nurses Includes; The traditional clinical training, crowded hospital wards and the density of other nurses, Mistakes in determining the type of patients, Lack of continuity in training days Lack of communication between nursing staff and faculty members and description of Nurse responsibilities in the patient's bedside is not Specified The challenges of clinical teaching from the standpoint of educators include; Lack of understanding of patients of the nursing profession, Inconsistency between the theoretical and practical training, conflict between educational objectives and expectations of training And expectations.

The Nature of Nursing Work in ICUs

The ICU environment portrayed in the literature is hectic and noisy and bombards people in it with all sorts of stimuli. As early as 1972, Hay and Oken described nursing in ICU as a demanding practice that poses high emotional and physical risks in providing care to patients who are very ill and often fighting for their life. Today, ICU nurses have to work faster and faster with sick people whose needs are becoming more and more complex in a setting in which technology is always advancing.

The empirical—mainly quantitative—studies reviewed describe the sources of dissatisfaction or stress at work of ICU nurses and report on the methods used to try to manage them. The studies also underscore the malaise produced by the difficult working conditions that stem in turn from the nursing shortage. These quantitative studies were generally conducted using self-administered questionnaires, and they tend to favor the stress-coping approach, which, Estryn-Béhard (1997) points out, calls on individuals' ability to adapt to the demands of the work environment by drawing on their personal resources and the support of people around them.

The Sources of ICU Nurses' Suffering At Work.

All the studies on mental health at work highlight the importance of the organization of work as a factor in suffering on the job. Suffering refers to the notion of psychological suffering and is related to the loss of pleasure, cooperation, solidarity, and social interaction. Gilbert defines workers' suffering due to the organization of work as "an extremely powerful force that is all too often ignored in interpreting behavior on and off the job". He contends that suffering emerges when, among other things, work loses its meaning, and that it sets in motion a struggle on the part of the workers to maintain their psychological well-being. The review of the literature provides some insight into various interrelated themes that are to varying degrees associated with the notion of stress. The particulars of the different analyses point to many signs of suffering among ICU nurses that can be deduced from the defensive strategies they employ.

Working as a nurse in ICU: a high-tension practice

Mark and Hagenmueller note that ICU nurses have to make a great many decisions in their practice and often have to deal with nursing situations that are characterized by high levels of complexity, uncertainty, instability, and variability. Bakker et al. The need for nurses to make many important decisions in their work in circumstances in which there is only limited latitude for decision making is an important stress factor for them. Similarly according to a qualitative study by Cronqvist, Lützén and Nyström, the dissonance created by these two contradictory elements- the duty to make decisions and the small latitude for making them- is a major source of stress for ICU nurses.

Demands of the technology

Grout offers an excellent analogy for the intensity of nursing in highly technologized ICUs in comparing it to the work of air-traffic controllers in busy airports. Similarly, a qualitative study by Alasad underscores the fact that new ICU nurses experience high levels of stress and fear before they achieve technological competence. Furthermore, some authors point out that nurses who are poorly prepared to use the technology suffer major stress and a feeling of overload that will lead them to focus dangerously on the equipment rather than on their patients.

The heavy workload of ICU nurses

In the early 1990s, the Quebec sociologist Carpentier-Roy used the comprehensive psychodynamics of work approach to investigate nurses' work in ICU. She reported that the sources of anxiety are more plentiful in that environment than in other units. The author observed that the nurses have a greater responsibility vis-à-vis their patients in ICU than in other units. They are thus subject to different sources of anxiety, such as the fear of not being able to use the various and growing number of technological devices properly; the anxiety arising from the impossibility of offering patients and their family the support they need when they need it; and the fear of not recognizing the signs calling for urgent intervention quickly enough. According to Carpentier-Roy, the greatest sources of suffering for ICU nurses are psychological, and the fact that they have to deal with death on a daily basis is a central feature of the suffering for them.

Continuous evaluation of the patients' clinical condition comprises such an important part of nursing in ICU that it is hard for nurses to deal with anything unexpected when it occurs, further adding to the cognitive and psychological burden of their work. The frequent interruptions demand additional intellectual effort on their part in their descriptive study, Robinson and Lewis report that the most important stressors ICU nurses identify include an atmosphere of crisis, which heightens psychological tensions, and the unavailability of an experienced physician in an emergency. In the frantic working environment of an ICU, in which the people under care are so highly dependent, nurses must stay calm and alert. They cannot hesitate or make a mistake, for a patient may die if they do. Furthermore, ICU nurses have to have a very high level of physical endurance because they spend extended periods on their feet and frequently have to move patients who are extremely heavy.

The perception of futile care

For ICU nurses, many work situations are felt to run counter to their values or vision of care. Meltzer and Huckabay note that there are many instances of what nurses perceive as aggressive therapy or nonbeneficial care and that there is a strong association between such perceptions and burnout among nurses. In a qualitative study of 25 ICU nurses in American hospitals, Badger concludes that, for them, the perception of futile care is the most important stress factor at work.

Lack of recognition at work

Contemporary studies often raise the issue of the lack of recognition at work. Carpentier-Roy observes that ICU nurses suffer more from lack of recognition for what they do and from the lack of power to match their responsibilities than from the actual physical demands of their work. Hay and Oken report that the lack of gratification from patients affects the self-esteem of ICU nurses. In an analysis of the severity of stress factors, Robinson and Lewis identify lack of recognition as one of the most important stressors that ICU nurses have to contend with.

 **Personnel shortages and instability of experienced nursing staff**

Budget restrictions and a lack of resources often limit nursing numbers and have thus made it hard to deal with the demands of care when there is a sudden deterioration in the condition of some ICU patients. This situation is quite common and is a major source of anxiety for ICU nurses. In many ICUs, the vacancy rate for positions on the evening and night shifts has tended to increase so much that it has become common practice to resort to mandatory overtime and independent personnel (agency nurses). Indeed, for lack of staff, ICU nurses must often work two shifts in a row; that is, 16 hours in all. The overtime they work and the need to continually readjust their private life gives rise to fatigue, anxiety, anguish, and frustration in nurses.

In the face of these harsh working conditions, many experienced nurses are quitting the ICUs and are taking with them the experience and expertise that acted as a safety net in the provision of care. The vacancies they leave are often hard to fill, and the consequent lack of nurses frequently makes it necessary to close ICU beds temporarily. By reducing accessibility to beds on these units, these circumstances aggravate emergency-department overcrowding and prolong waiting times for major surgery that requires a stay in ICU. The situation thus negatively affects the performance of the entire Quebec healthcare system.

Too short an integration period for new nurses

The lack of nurses is so critical that, it has been observed, most ICUs no longer set eligibility criteria for nurses wanting to practise in them. However, clinical situations encountered in ICU are so complex they demand specific technical, technological and relational nursing skills as well as a finely honed clinical judgment that are acquired principally from working in the ICU environment and from the support of experienced peers. Most of the experienced nurses who choose to stay in ICU are therefore continually obliged to act as preceptors to the new nurses in the hope that the novices quickly become functional enough to pull their weight on the team. Thus, in addition to having to bear the loss of colleagues who have left the unit, nurses who remain on the job have to provide both clinical and psychological coaching and support to new nurses. Furthermore, Alspash stresses the fact that the time allotted to integrating new recruits into the ICU is often too short to allow them to appropriate the role of intensive-care nurse. Their psychological tension rises, and they feel very insecure when, after too short a period of integration and without much support, they have to assume overall management of two patients with complex needs. Many of them feel overwhelmed and decide to leave the ICU for other units. The circumstances thus contribute to the very high turnover rate in ICUs and the heavier workload for the expert nurses who stay on the job and alone have to shoulder responsibility for providing all the care that patients on the unit require.

To sum up, the sources of suffering at work of ICU nurses that have been identified in the literature foreshadow danger for their mental health. ICU nurses experience a variety of situations and an organization of work that produce feelings ranging from slight discomfort to profound suffering. We shall now turn to the methods they use to protect themselves from such suffering.

Defensive strategies used by ICU nurses

According to Dejours, defensive strategies are behaviors workers use to lessen the perception of suffering. These mechanisms are not pathological per se according to Alderson; they are the very expression of what is problematic in the contemporary organization of nursing practice. Their purpose is to minimize perceived suffering rather than suppress it at its source, and they are implemented by workers in reaction to a restrictive organization of work and to work that no longer has meaning. The literature indicates that ICU nurses employ defense mechanisms to protect themselves from the suffering they have experienced.

It should first be noted that a great number of descriptive empirical articles report that ICU nurses use a variety of coping strategies to adapt to stressful situations. They may accordingly externalize their feelings to the group or engage in relaxation activities such as listening to music, doing aerobic exercise, watching television, walking, reading, seeking solitude, and going on vacation.

Secondly, Hay and Oken report in their descriptive study that ICU nurses are usually more detached from their work than nurses in general wards because the ICU nurses use defense mechanisms that mitigate their perception of suffering. The mechanisms include denial of situations that are sources of suffering (for example, denying the anxiety associated with continual exposure to death); avoidance of the emotional impact of the work (for example, focusing on technical tasks rather than on the human aspect of care); and absenteeism (for example, taking a day off after a trying experience at work). In a comparative study, Maloney observes that ICU nurses rationalize their feelings to protect themselves from difficult experiences on the job. In a recent qualitative study, Badger (2005) reports that ICU nurses use behavioural coping mechanisms in addition to the usual emotional and cognitive ones. These behavioural mechanisms include withdrawal, avoidance and distancing oneself on the part of the nurse.

Thirdly, Carpentier-Roy reports another common strategy used by ICU nurses: "the ideology of the elite". When an ICU nurse made statements stressing the high value attached to danger as a source of pleasure, the author noted that ICU nurses call on this elite ideology to protect them from the anxiety produced by psychological factors. The ideology serves as a defence strategy by reminding them that not everyone can be an intensive care nurse. According to Carpentier-Roy, it thus helps ICU nurses "bear the unbearable" by protecting them from the anxiety they are subject to.

Our review of the literature thus reveals that ICU nurses resort to a range of adaptation mechanisms that are appropriate when viewed in terms of a stress-coping approach or as defensive strategies undertaken to ease the perception of suffering or to resist assaults on their emotional well-being. As Carpentier-Roy says, in ICUs there seem to be ways to release the often unavoidable psychological tensions that nurses experience. According to the PDW approach, individual nurses are likely to find such channels, which Alderson calls "outlets," in their sources of pleasure at work.

According to this, the fields of clinical education that considered as a field of challenging In fact are the philosophy elements of clinical nursing education. Despite the fairly wide reporting in the literature of the many roles of clinical supervision by the nursing teacher, little attention has been given to conceptualizing the relative priorities these roles take during the process of supervising nurses in clinical practice (Oerman, Garvin, 2002). Another the effective factors in clinical education of nurses are the clinical instructors ,As a result, all teachers were assigned to clinical areas in which they were to facilitate the development of an educational milieu to support nurse learning in practice. .Nurse educators face many challenges in the current healthcare environment. Educational methods, philosophies, and the content of curricula need to be re-examined to meet the needs of professional nurses who will practice in the next millennium. Evidence-based nursing is one approach that may enable future healthcare providers to manage the explosion of new literature and technology and ultimately may result in improved patient outcomes.

Nurse's perception teaching revisited despite a wealth of research on clinical teaching, the criteria for determining what constitutes effective clinical teaching remain poorly defined (Sook W, Cholowski K, Williams AK 2002). Nurses' and clinical educators' perception of characteristics of effective clinical educations in school of nursing.

As new graduates enter the workplace, they are faced with many stresses associated with beginning practice. The nurse manager, preceptor, and nursing staff need to be aware of the specific stresses and challenges experienced by new graduates during their orientation period and need to plan interventions for coping with them(Oerman & Garvin 2002).

Challenges in the healthcare environment are forcing healthcare organizations to examine new practice models to reduce costs while maintaining quality of care. To respond to the changes in the practice environment, organizations can alter their practice arrangements. Nursing practice models are innovative practice arrangements that differ from traditional models on one or more of the following structural dimensions:

- The degree to which the practice of individual nurses is differentiated according to education level or performance competencies;

- The degree to which nursing practice at the unit level is self-managed, rather than managed by traditional supervisors;

- The degree to which case management is employed; and

- The degree to which teams (either nursing or multidisciplinary) are employed. Many practice models contain more than one of these elements and also include elements of primary nursing (Weisman, 2007).

Other structural dimensions may be used, but overall, some practice models are intended to optimize costs, while others are intended to deal with staffing constraints. Policy makers should support evaluation and adoption of innovative models that are intended to address challenges facing the nurse practice environment. Such models should provide learning opportunities that emphasize safe, coordinated, and affordable healthcare

Nurses encounter some challenges in dealing with clinical learning environment and in interaction with instructors, patients, and department personnel. Many nurses stated that they had the most interactions with the instructors and believed that the way an instructor treats a nurse affects their exposure to clinical learning environment. One nurse stated the following. In addition to the improper treatment of instructors toward the nurses, some behaviors of nurses are also oppressive to nurses. One of the nurses described how improperly the department nurse treated her.

Also, inadequate knowledge, deficient practical skills, and insufficiently developed communication skills; Many nurses did not have sufficient knowledge to care at the bedside when dealing with clinical learning environment and providing care to the patients was challenging for them. One of the nurses said the following. Clinical environment is a suitable context for learning skills needed to care for patients. However, some of them are considered basic health care skills and any deficit in them affects the quality of care. In this regard, nurses had difficulties in performing procedures in some situations, due to the lack of necessary skills. Deficiency in practical skills in caring for patients was a concern of many nurses in the clinical setting.

Many nurses mentioned the lack of communication skills as the reason for deficiency in communicating with the clinical learning environment. Insufficiently developed communication skills sometimes cause disruption in providing care for patients. Many of the nurses participating in this study became distressed and overwhelmed in dealing with new experiences within the clinical learning environment. From the perspective of these nurses, providing care for patients is stressful to them. Inferiority complex was more evident among female nurses than male ones. Clinical teaching is carried out by both male and female clinical teachers who liaise closely with the course coordinator. They are required to have completed the university undergraduate nursing degree program and at least 3 year of post-registration clinical experience. Nursing lecturers (those who teach the theoretical component of the course) may hold a master or doctoral degree in nursing and should have a good clinical experience. Nurses are assigned to different clinical teachers for each clinical placement throughout the semester.

Graduating clinically competent nurses is probably the principal objective of curricula and clinical teachers. Nursing education consists of theoretical knowledge combined with clinical practice. Nurses require more than the traditional theoretical classroom teaching as there is so much in the nursing field that is best learned by doing not just talking. Clinical practice increases nurses' knowledge, and their capacity to synthesize theoretical knowledge and nursing care according to Addis and Karadag in 2003. Also, the main factor in the development of nursing capabilities is close observation of clinical practice. In line with Florence Nightingales ideals, clinical teaching that is teaching and learning focused on, and usually directly involving patients and their problems lies at the heart of nursing education. Clinical teaching in nursing has been defined as the mode that provides nurses with the opportunity to translate theoretical knowledge into the learning of a variety of skills required to give patient-centered care. The main aim of clinical education is to provide nurses with necessary competencies in both nursing and clinical skills.

Clinical Nursing Challenges

Clinical teaching internationally is seen internationally as an important part of nursing education. The literature suggests that clinical learning is affected by many factors, including the quality of supervision and feedback, and characteristics of learners and teachers. Nursing education receives surprisingly little attention from the nurse educators. It suffers from the lack of coherent theoretical base necessary to inform participants. There is also lack of substantial research in the area of clinical teaching which is the 'heart' of nurses' professional nursing education. The clinical teaching in lacks effectiveness which indicates a need for more active clinical setting to be able to make the theoretical components come alive in the practice and enthuse nurses.

Teaching in the clinical environment in has many challenges such as time pressure which is everyone's enemy. Also, there are potential problems and deficits which might act as a barrier toward achieving a good environment for clinical teaching. These barriers can be divided into two parts; some are related to the clinical teachers while the others are related to the nurses. Problems encountered by clinical teachers include lack of clear objectives and expectations and inadequate supervision and provision of feedback. Nurses may have little opportunity for reflection and discussion and lack of congruence or continuity with the curriculum.

A further problem is that the number of clinical teachers in the clinical areas is less than required compared to the number of nurses. The rapid increase of admission of nurses of one of the nursing schools in which cannot be met with a limited number of clinical preceptors.

Furthermore, Most of the clinical preceptors lack the competence to employ effective learning strategies by being able to employ teaching methods, display solid communication skills, and have the ability to motivate nurses which have impacted negatively in the teaching process. The lack of adequate preparation of clinical instructors before they embark on teaching impacts greatly on the teaching learning process. Clinical educators must know what kinds of experiences facilitate or hinder the learning process. The educator's selection of learning theories and structure of the learning experience become important in this realm. To be effective, educators must have knowledge of materials to be learned, the learner, the social context, and educational psychology.

With respect to learning, learners' readiness has direct effect on the teaching process. This can inhabit nurses' enthusiasm to participate actively in the learning process. Readiness to learn can be defined as 'the time when the learner demonstrates an interest in learning the type or degree of information necessary to become more skilful in job. Therefore, the lack of the nurses' enthusiasm prevents them from the participation in the teaching process. Moreover, clinical area is a place of multiple languages. With English as the language of communication, Saudi nurses find it hard to communicate using English language in open and busy areas like hospitals. Despite the fact that the entirely nursing curriculum program is instructed by English language nurses find it difficult to communicate in English. Nurses prior to college enrollment do not have sufficient exposure to the English language and English is not taught intensively to enable them to use it in sophisticated manner later in workplace. Accordingly, the process of transferring, understanding and receiving information among nurses and other hospital workers is being affected due to the English language proficiency of many nurses.

 # SOLUTIONS TO OVERCOME CHALLENGES OF CLINICAL TEACHING

Based on the results of many study, many nurses lack the communication skills necessary for effective communication in the clinical environment. It is suggested that the effective communication skills are taught to nurses before they enter the clinical environment with the emphasis on the differences between the clinical environment and the classroom environment. many nurses mentioned lack of theoretical knowledge and practical skills as one of the problems involved in caregiving. Therefore, before nurses enter the clinical environment, it should be ascertained that they are theoretically and practically prepared as they take tests and give care in the skill lab. In light of the presence of stress and inferiority complexes in nurses in confronting the clinical environment, it is suggested that while they receive psychological consultation on the nursing profession, caregiving, and the hospital environment plans be made for them to visit the hospital and to get acquainted with the clinical learning environment before they begin the actual internship.

It is one of the teachers' major responsibilities to treat nurses properly in the clinic, causing higher enthusiasm and motivation for learning as well as increasing their self-confidence. Nabolsi et al. demonstrated in their study that proper treatment and establishment of a communication with nurses are an important item for nursing teachers to be a role model for nurses. Training that involves value and respect facilitates the teaching-learning process and socializes the nurses into the nursing profession. Authors found that proper communication with nurses increased their motivation.

In Baraz-Pordanjani et al.'s study, discrimination in the use of educational facilities and amenities and also in interpersonal communication was reported as a factor distorting the nurses' professional identity in the clinic, which is in line with the results of our study. The comparison between nursing and medicine and regarding medicine as a superior major violates nurses' personal dignity and gives them a sense of professional inferiority. Nurses' inadequate preparation for entering the clinical environment creates problems for them and nursing teachers. Even though they learn the fundamentals of nursing in classrooms and practice rooms, nurses do not have sufficient time to practice and repeat these skills to completely enter the clinic. Killam and Heerschap found that the nurses' insufficient practice and lack of skill before entering the clinical environment created problems for them with respect to learning in the clinic. Moreover, the nurses' lack of skill in confronting the clinical environment and dealing with actual patients is evident. Nurses' lack of knowledge and skill and inadequate preparation for entering the clinical environment disturb their learning processes and make them anxious. Acquisition of communication skills in nurses creates a guiding atmosphere in the clinical environment, followed by an increase in their motivation. Nurses' lack of practical skills is considered as a challenge in entering the clinical environment.

Nurses' stress in confronting the clinical environment affects their general health and disturbs their learning processes. According to one study, stress is one of these nurses' experiences in the clinical environment. In Changiz et al.'s study, it was revealed that the causes of nurses' stress in the clinical environment fall into three types of stress due to the educational plan, stress due to the educational environment, and factors concerning the nurses. In Chesser-Smyth's study, stress and anxiety were one of the nurses' experiences in the clinical environment. Nurses' young age when entering the clinical environment and their social and emotional lack of experience lead to stress and psychological problems.

An inferiority complex is another challenge mentioned by the nurses participating in the study. The results of Edwards et al.'s study showed that low self-confidence is one of the nurses' problems. Adequate self-confidence is one of the nurses' requirements in providing good care. In Joolaee et al.'s study, lack of self-confidence has been referred to as a major cause of fear and anxiety in nurses. The researchers demonstrated that lack of self-confidence also disturbs communication in nurses. Moreover, having adequate self-confidence for caregiving is one of the most important factors affecting the nurses' learning. In Begley and White's study, self-confidence was an important part of a nurse's personal and professional identity. We found that nurses are faced with many challenges in the clinical learning environment, which affect their professionalization and learning processes. Many nurses are not mentally prepared to enter the clinical environment leading to higher rates of psychological problems. Moreover, lack of adequate knowledge and skill along with lack of mental and psychological preparation disturbs the learning and patient caregiving processes. Improper treatment, discrimination, inadequate knowledge and skill, and lack of communication skills in these patients lead to stress and inferiority complexes in them. In view of the nurses' challenges in confrontation with the clinical learning environment and the necessity of learning and providing patients with care in a peaceful environment free of any tension, educational authorities and nursing faculties are required to pay particular attention to these issues and try to facilitate the nurses' learning and professionalization.

 ### Teaching in the ICU: From Behaviors to Boards to Barriers

Educators seeking to influence their learners want to know how to teach in the ICU and how to equip their learners for board certification. One multicenter survey study focused on what specific teaching behaviors internal medicine residents valued in ICU attendings. They found that residents valued attendings who enjoyed teaching, expressed empathy and compassion to patients and families, explained clinical reasoning and differential diagnoses, treated nonphysician staff members respectfully, and demonstrated enthusiasm on ICU rounds.

These teaching tactics can be refined and continuously improved, making them ideal for faculty development. Beyond behaviors that educators can learn, they also need to know the targets for board certification to prepare their learners. A multidisciplinary panel of critical care education experts in the American College of Critical Care Medicine established guidelines on the specific credentials (eg, Advanced Cardiac Life Support), cognitive skills (eg, management of shock and myocardial infarction), and procedural skills (eg, central line insertion) that critical care trainees should master. However, despite the comprehensive nature of this report in describing skills and diagnoses that trainees must master, little guidance was given on educational methods; the article broadly stated that critical care trainees should master these skills through "the use of any of a number of techniques, including didactic lectures, journal club sessions, and illustrative case reports."

Informal case-based teaching sessions (91%) or didactic lectures (75%). Although bedside teaching was the most frequent teaching method used in the ICU, there is no clear standardized approach to critical care education.

However, there are barriers to consistent and effective bedside teaching in the ICU, related to both faculty and trainees. Faculty barriers include clinical workload and lack of protected time and funding, as well as the Accreditation Council for Graduate Medical Education regulations, which have consequences on resident continuity of care, attitudes, and availability for teaching. Interestingly, 63% of respondents believed that faculty had changed their approach to teaching in the ICU as a result of trainee duty hour restrictions. In the face of these barriers, how can faculty and trainees maximize their teaching while in the busy ICU environment? The next sections focus on practical evidence-based tips and tricks to improve teaching in the ICU.

Although bedside teaching was the most frequent teaching method used in the ICU, there is no clear standardized approach to critical care education. However, there are barriers to consistent and effective bedside teaching in the ICU, related to both faculty and trainees. Faculty barriers include clinical workload and lack of protected time and funding, as well as the Accreditation Council for Graduate Medical Education regulations, which have consequences on resident continuity of care, attitudes, and availability for teaching. Interestingly, 63% of respondents believed that faculty had changed their approach to teaching in the ICU as a result of trainee duty hour restrictions.

In the face of these barriers, how can faculty and trainees maximize their teaching while in the busy ICU environment? The next sections focus on practical evidence-based tips and tricks to improve teaching in the ICU.

 **Teaching under Time Pressure**

The ICU clinical environment is extremely fast-paced, with the possibility of multiple patients decompensating at the same time; teaching, therefore, has to be time-efficient and high-yield. Previous studies have found that time pressures and high clinical workloads were the most significant barriers for faculty to change their teaching styles in the ICU. Fortunately, there is robust medical educational literature from the outpatient, ED, and inpatient settings discussing how to approach teaching in a time crunch.

Within minutes, the teacher can quickly assess the needs of individual learners by asking questions, selecting a model for rapid teaching such as the "1-minute preceptor" model, and providing quick constructive and corrective feedback on performance. This model was originally based on the five-step "microskills" framework for clinical teaching initially developed for outpatient clinic-based faculty precepting: it encourages teachers to get a commitment from the learner, probe for supporting evidence, teach general rules, reinforce correct behaviors, and correct errors.

Mini-chalk talks are another useful method of quick, efficient bedside teaching that eschew lengthy PowerPoint presentations in favor of brief, visual, on-the-fly teaching moments. On busy ICU rounds, preselecting which patients would be best for the focus of certain bedside teaching points is an effective and practical way to incorporate teaching in a time-efficient manner. Preselecting patients for high-yield teaching is frequently done in the outpatient setting.

- Use case examples recently seen on-service to build up a "bank" of mini-chalk talk topics.
- Take time to practice and hone the talk in advance, but be flexible to accommodate learners' questions – purely ad lib talks can be unfocused.
- Use simple, high-yield figures, such as Venn diagrams, simple tables, flow charts, graphs, timelines, etc, instead of PowerPoint slides.
- Consider "gamification" of the teaching session to encourage participation (eg, set up a competition to see what team can assemble a thoracentesis kit accurately in the shortest amount of time).
- Consider holding your mini-chalk talk outside of a traditional conference room (eg, at the bedside, outside the patient's room on rounds, etc).
- Avoid the temptation to over-teach – multiple short high-yield teaching sessions about different aspects of the same

talk may lead to better retention than one long PowerPoint presentation this technique can be effectively applied in the ICU to save time and focus bedside teaching..

- Teaching Tips for the Bedside and ICU Rounds "Twelve tips to improve bedside teaching" is the title of a concise and comprehensive article that describes key strategies involved in bedside teaching prior rounds, and following rounds.

Teaching Tips for the Bedside and ICU Rounds

Prior to rounds

1. Preparation: review the curriculum, know the teaching audience, improve their own subject knowledge, and consider further training in education

2. Planning: outline clear goals/expectations for the teaching encounter

3. Orientation: orient learners with session plan and goals/objectives for the session

During rounds

4. Introduction: Introduce team to patient and explicitly state that this moment is a teaching encounter

5. Interaction: Treat patient as a role model of an ideal physician-patient interaction

6. Observation: Keenly observe learners' interaction with the patient to learn more about modifying the session

7. Instruction: Engage the learner in the teachable moment and gently correct if errors are made

8. Summarization: Summarize teaching points encounter

Following rounds

9. Debriefing: Explicitly leave time to debrief outside the room to answer questions, raise more questions, assign readings

10. Feedback: Build-in time for feedback about the teaching encounter itself

11. Reflection: Personally reflect on what went well and what could be improved upon for next time

12. Preparation: Use insights from reflection to prepare and improve next session rounds, the teacher is encouraged to build-in time for debriefing, feedback, and reflection, using these to prepare for the next session.

The published literature on teaching techniques specific to the ICU is more limited. Given the time barriers we have previously discussed, ICU bedside teaching has to be deliberately and carefully refined. Educators should thoughtfully consider the limited scope of a bedside teaching session and resist the temptation to over-teach. Instead, giving a brief 5-min to 15-min talk at the bedside that is relevant to current or recently admitted ICU patients may be more helpful.

Carlos et al describe a useful framework to discuss how to incorporate ICU teaching in minimal time at the bedside: the CARE framework focuses on learning climate, attention to the teaching encounter, critical reasoning, and learner evaluation.

Climate (C)

    Set learner expectations and be explicit observation

    Avoid medical jargon and one-upmanship

    Explain purpose of encounter and encourage participation

    Set patient expectations and ask permission

Attention (A)

    Plan encounter in advance

    Remain focused in the moment

    Keep content relevant for all learners

    Maintain democratic leadership style

Reasoning (R)

    Encourage hypothesis-driven examination

    Ask probing questions

    Avoid "read-my-mind: questions

    Give formative feedback focused on behaviors

Evaluation (E)

    Avoid pointed criticism

    Encourage reflection after the encounter

    Compile observations for summative feedback

The CARE framework and the critical reasoning strategies provide learner-focused practical tips for ICU trainees and faculty alike to incorporate into ICU bedside teaching.

 **Critical Components of Crew Resource Management with (Skills and Examples)**

Situational awareness: Active involvement of all team members to visualize and acknowledge the field:

Skill: Closed-loop communication helps incorporate the perspectives of all team members and allows verbal "teach-back" of actions

Example: Physician: "Let's set a goal for this patient to get out of bed today." Physical Therapist: "Agreed, I'll make it a goal to at least mobilize the patient out of bed to chair once today." Physician: "Yes, thank you for clarifying." Problem identification: Use of voluntary, active, and open communication to declare concerns.

Skill: Shared mental models help ensure all team members are on the same page about the problem

Example: "At this point, I'd like to summarize where we are with this patient and get your feedback and perspectives before we talk to the family again Decision-making: Generation of alternative acceptable solutions through accurate anticipation and diagnosis of problems.

Skill: Inviting consensus helps rely on the expertise of multiple team members to ensure that diverse opinions are considered to avoid premature closure

Example: "Team, I'd like each one of you to weigh in on whether or not the patient is ready for extubation." Workload distribution: Assignment of tasks so that no team member is overloaded.

Skill: Intentional delegation and communication of who performs specific tasks

Example: "As our ICU Team Pharmacist, could you please look up whether the new antibiotic could be associated with the abnormal liver function tests?"

Teaching ICU Knowledge and Skills Teaching Team Management

The ICU team is diverse and interdisciplinary. It may include various professionals such as pharmacy, nursing, and respiratory therapy, as well as differing learner levels from nurses to fellows. Lessons in communication have been learned from the airline industry. The aviation Crew Resource Management curriculum promotes cooperation and coordination of skills that parallels health care models. This model has been increasingly used in the quality improvement literature in health care, but it has been explored in less detail in medical education (outside of the simulation literature).

Key components to effective bedside teaching with a team include orienting all involved to roles, responsibilities, and expectations. Open communication should be encouraged, and educators should strive to create a learning environment that is safe and fosters inquisitive discussion and critical thinking.

Finally, interprofessional education can occur outside the hospital as well. Several courses that include simulation have been developed to train teams together. Ensuring psychological safety of all interprofessional team members is critical to ensure high-performing teams, especially during high-stakes situations such as resuscitations and codes.

Teaching Patient and Family Communication Skills

The ICU The literature on bedside teaching in the ICU has often focused on the transmission of medical knowledge, but ICU teaching is also particularly unique for its focus on communication, particularly at the end of life. Teaching about communication, specifically in the context of family meetings in the ICU, is another invaluable skill for trainees and a teaching opportunity for faculty and fellows. Miller et al argue that communication in the ICU should be taught similarly to invasive procedures (eg, central line insertions) by focusing on training, observation, and feedback on communication skills to achieve mastery.

Different studies provide a series of tips focusing on communication skills for difficult conversations in a clinical situation, which are particularly applicable to the ICU setting. They recommend assessing the learners' baseline communication skills as well as their understanding of the clinical situation and then matching the learners' educational needs with their assigned roles in the difficult conversation. Prior to a family meeting, the meeting agenda should be reviewed with the learner; the instructor should prepare for the meeting by discussing the learner's individual goals and reviewing pertinent communication skills ahead of time. Once in the meeting itself, the patient and family should be informed that the trainee will lead the meeting, and the instructor should actively observe and take notes during the encounter. It is important to avoid interruptions yet be prepared to step in if needed. Following the encounter, the learner should be debriefed, providing opportunities for reflection and corrective feedback, while making an action plan for the future.

Teaching Procedural Skills

Teaching procedures in the ICU is also an important part of the educational experience. A six-step framework for teaching of procedural skills can help the instructors best communicate the information. First, learners have to acquire the cognitive knowledge about the procedure through learning about it in didactic lectures and observing the procedure itself. The learner then ideally should practice the procedure using simulation and prove competency prior to performing the procedure on a patient. The learner then graduates to performing the procedure on the patient while being observed by the teacher. Continuing practice helps maintain these skills. This "learn, see, practice, prove, do, maintain" framework is easy to implement even in the busy ICU and goes beyond the simple "see one, do one, teach one" model.

 **Six-Step Framework for Procedural Skills Teaching**

1. Learn: Learn about the procedure via reading or videos.
Example: "Please review the New England Journal of Medicine video on thoracentesis before this afternoon's procedure."

2. See: Instructor demonstrates the skill, first nonverbally, then going through the individual steps with verbal description.
Example: "First I'm going to show you how to do the thoracentesis on the mannequin simulator, then I'm going to show you again, this time, talking you through each step."

3. Practice: Deliberate practice on simulator
Example: "For this next step, I'd like you to really focus on keeping the needle perpendicular to the patient's body."

4. Prove: Summative assessment and feedback on simulator
Example: "Now I'm going to watch you do an entire thoracentesis on the mannequin simulator and give you feedback afterwards."

5. Do: Direct supervision of performance on human
Example: "I'm going to directly supervise you doing your first thoracentesis on the patient."

6. Maintain: Maintenance of skill through clinical practice supplemented by simulation as needed
Example: "It's been over a year since your last thoracentesis, so let's return to the simulation lab to refresh ourselves quickly."

Learn, see, practice, prove, do, and maintain: an evidence-based pedagogical framework for procedural skill training in medicine.

Some training programs have adopted required communication training during ICU rotations and have found improvement in residents' perceived skills and positive family member responses. Bhang and Iregui recommend that family meetings be structured on the foundation of a relationship of trust, framed by the patient and family explanatory models as well as the medical facts, and supported by collaborative decision-making.

Reviewing pertinent communication skills ahead of time. Once in the meeting itself, the patient and family should be informed that the trainee will lead the meeting, and the instructor should actively observe and take notes during the encounter. It is important to avoid interruptions yet be prepared to step in if needed. Following the encounter, the learner should be debriefed, providing opportunities for reflection and corrective feedback, while making an action plan for the future.

 **Case presentation**

"Working in the ICU was one of the most beneficial experiences. On first day, I was assigned to a helpful nurse (we'll call her Nurse C) who had been in ICU for 27 years. Admittedly, I was a little intimidated at first because telemetry is not my forte! Seeing patients connected to multiple tubes and machines – not to mention all of those beeping sounds – did nothing to ease my nervousness, but I was determined to overcome my fear and gain confidence as a new nurse. Nurse C asked me to tell her a little bit about my background, and we discussed my strengths and weaknesses, and things I would like to work on. She also asked if I would be willing to perform the entire care for one of her patients, under her supervision, of course. So, the day begins. After we did rounds on all three of her patients, Nurse C assigns me to a patient on a ventilator, in congestive heart failure, on telemetry with a foley and a PEG. I did the head-to-toe assessment as thoroughly as I could since ICU patients require a more in-depth assessment. I asked Nurse C to guide me through it because I was not yet confident enough to do the cranial nerves assessment on my own. She was thankfully more than willing to do so, and explained everything in detail. Medication pass for my patient was due and she quizzed me on the use, side effects and labs to watch out for on each. Things were going pretty smoothly. ICU clinical was absolutely one of the best learning experiences I've had had in my nursing education. In one day I was able to overcome my fear of working on a telemetry floor, accurately read my patient's telemetry strips, understand from experience what a high pressure ventilator alarm meant, and practice deep suction. I am so grateful that I found the courage to practice skills I had learned in theory, and was in turn able to feel confident practicing what deep down, I already knew I could do and I was finally thinking like an RN, I began to understand that I am prepared and that I can do this! And this was just after Day One."

 **CONCLUSION**

Excellent clinical teaching is a skill that can be studied, refined, and continuously improved, just as any other procedure. Teaching in the ICU comes with unique challenges given the medical complexity of the patients, the time pressure, the diverse levels and professions of learners, and the challenges of communication at the end of life. In the present book, teaching tips transcend professional barriers: and use the term "learner" broadly to encompass all the aforementioned types of interdisciplinary learners.

ICU bedside teaching is a formative experience for physicians and professionals. The ICU environment is conducive to teaching critical thinking skills and demonstrating key communication skills such as empathy. For bedside teaching to remain valuable, we encourage educators to pay attention to details throughout the rounding process (prior to, during, and following rounds). Educators frequently perceive that time is the biggest barrier to ICU teaching; we therefore have highlighted tactics for teaching "on the fly" that are high value, such as the 1-minute preceptor and use of mini-lectures. In the ICU specifically, the CARE teaching framework focuses on learning climate, attention to the teaching encounter, critical reasoning, and learner evaluation. These proven successful strategies will hopefully inspire instructors to practice continuous quality improvement of their teaching skills.

Notes

References:

- A. Chesser-Smyth, "The lived experiences of general nurse nurses on their first clinical placement: a phenomenological study," Nurse Education in Practice, vol. 5, no. 6, pp. 320–327, 2005.

- A. Killam and C. Heerschap, "Challenges to nurse learning in the clinical setting: a qualitative descriptive study," Nurse Education Today, vol. 33, no. 6, pp. 684–691, 2013.

- A. Krueger and M. A. Casey, Focus Groups: A Practical Guide for Applied Research, Sage, Thousand Oaks, Calif, USA, 2009.

- Addis G, Karadag A (2003) An evaluation of nurses' clinical teaching role in Turkey. Nurse Educ Today 23: 27-33.

- Alasad J (2002) Managing technology in the intensive care unit: the nurses' experience. International Journal of Nursing Studies 39: 407-413.

- Alderson M (2004c) La psychodynamique du travail et le paradigme du stress: une saine et utile complémentarité en faveur du développement des connaissances dans le champs de la santé mentale au travail. Santé mentale au Québec 29: 261-280.

- Alderson M (2005) La souffrance psychique des infirmières. Est-ce pertinent de l'investiguer au moyen de la psychodynamique du travail?. Frontières 17: 53-58.

- Al-Omrani A. (2008) Perceptions and attitudes of

Saudi EFL and ESL nurses toward native and non-native English-speaking teachers (Doctoral dissertation). Retrieved from ProQuest Dissertations and Theses database. (Publication No. 3303340)

- Alspach JG (2003) Recognizing and Rewarding Nurse Preceptors in Critical Care. Critical Care Nurses 23: 13-19.

- American Association of Critical Care Nurses (AACN) (2005) Standards for establishing and sustaining Healthy work environments.

- B. Gaberson, M. H. Oermann, and T. Shellenbarger, Clinical Teaching Strategies in Nursing, Springer, New York, NY, USA, 2014.

- Badger JM (2005) A descriptive study of coping strategies used by Medical Intensive Care Unit nurses during transitions from cure- to comfort-oriented care. Heart & Lung 34: 63-68.

- Bakker AB, Le Blanc PM, Schaufeli WB (2005) Burnout contagion among intensive care nurses. Journal of Advanced Nursing 51: 276-287.

- Baraz-Pordanjani, R. Memarian, and Z. Vanaki, "Damaged professional identity as a barrier to Iranian nurses' clinical learning: a qualitative study," Journal of Clinical Nursing and Midwifery, vol. 3, no. 3, pp. 1–15, 2014.

- BasavanthappaB. T. (2003) Nursing education. Jaypee Brothers

- Bastable S. B. (2008) Nurse as educator. Jones &

Bartlett Publishers.

- Beckmann U, Gillies DM. (2001) Factors associated with reintubation in intensive care: an analysis of causes and outcomes. Chest 120: 538-42.

- Bourbonnais, R., Malenfant, R., Viens, C., Vézina, M., Brisson, C., Laliberté, D., et al. (2000).

- Boyle DK, Bott MJ, Hansen HE, Woods CQ, Taunton RL (1999) Managers' leadership and critical care nurses' intent to stay. American Journal of Critical Care 8: 361-371.

- Brown, L. O'Mara, M. Hunsberger et al., "Professional confidence in baccalaureate nurses,"Nurse Education in Practice, vol. 3, no. 3, pp. 163–170, 2003.

- Buerhaus PI, Staiger DO, & Auerbach DI (2000) Why are shortages of hospital RNs concentrated in specialty care units? Nursing Economics 18: 111-116.

- Burke R J (1993) Organizational-level interventions to reduce occupational stressors. Work and Stress 7: 77-87.

- Carpentier-Roy M-C (1990) Organisation du travail et santé mentale chez les infirmières en milieu hospitalier, Unpublished doctoral thesis, Université de Montréal, Département de sociologie, Faculté des arts et des sciences.

- Carpentier-Roy M-C (2000) Être reconnu au travail : nécessité ou privilège? In Actes du colloque Travail, reconnaissance et dignité humaine, Montreal.

- Carpentier-Roy M-C (2007) Reconnaissance au Travail : Un élément essential de plaisir et d'efficacité au travail. Lecture given at the Hôpital Maisonneuve- Rosemont during the Semaine de la reconnaissance du C. H., Montréal.

- Cavalheiro AM, Moura Junior DF, & Lopes, AC (2008) Stress in nurses working in intensive care units. Rev Lat Am Enfermagem 16: 29-35.

- **Changiz, A. Malekpour, and A. Zargham-Boroujeni, "Stressors in clinical nursing education in Iran: a systematic review," Iranian Journal of Nursing and Midwifery Research, vol. 17, no. 6, article 399, 2012.**

- Collière MF (2001) Soigner... Le premier art de la vie. Second edition. Paris: Masson.

- Conseil International des Infirmières (2006) La pénurie mondiale de personnel infirmier : domaines d'action prioritaire, Geneva, Switzerland.

- Conseil International des Infirmières (CII) (2004) La pénurie mondiale d'infirmières diplômées- aperçu des questions et des solutions: Initiative mondiale pour la révision des soins infirmiers, Geneva, Switzerland.

- Corley MC, Elswick R K, Gorman M, Clor T (1995) Moral distress of critical care nurses. American Journal of Critical Care 4: 280-285.

- Cronqvist A, Burns T, Theorell T, Lützen K (2004)

Caring about- Caring for: moral obligation and work responsibilities in intensive nursing. Nursing Ethics 11: 63-76.

- Cronqvist A, Lützén K, Nyström M (2006) Nurses' lived experiences of moral stress support in the intensive care context. Journal of Nursing Management 14: 405-413.

- D. F. Polit and C. T. Beck, Essentials of Nursing Research: Appraising Evidence for Nursing Practice, Lippincott Williams & Wilkins, Philadelphia, Pa, USA, 2013.

- Deasy, O. Doody, and D. Tuohy, "An exploratory study of role transition from nurse to registered nurse (general, mental health and intellectual disability) in Ireland," Nurse Education in Practice, vol. 11, no. 2, pp. 109–113, 2011.

- Dejours C (1999) Incidences psychologiques des nouvelles formes d'organisation du travail, du management et de la gestion des entreprises. Archives des maladies professionnelles et de médecine du travail, 60: 533-541.

- Diehl-Oplinger L, Kaminski MF (2001) Need critical care nurses? Inquire within. Dimensions of Critical Care Nursing 20: 30-32.

- Dracup K, Bryan-Brown CW (1998) One more critical care nursing shortage. American Journal of Critical Care 7: 81-83.

- E. Manninen, "Changes in nurses' perceptions of nursing as they progress through their education," Journal of Advanced Nursing, vol. 27, no. 2, pp. 390–398, 1998.

- Edwards, P. Burnard, K. Bennett, and U. Hebden, "A longitudinal study of stress and self-esteem in nurse nurses," Nurse Education Today, vol. 30, no. 1, pp. 78–84, 2010.

- Elhabashy, S., BRONCHIAL HYGIENE THERAPY: Modalities & Techniques , , New Jersey, Princeton University Press, 2016.

- Elhabashy, S., Cardio-Thoracic Injury, Essentials All Critical Care Nurses Need To Know, U.S.A, Elseveir (MOSBY), 2015. the_final_book.pdf

- Elhabashy, S., Clinical Alarms Hazards and Management at Critical Care Settings, , Newyork, LULU Publishing Com., 2015. 16932240_cover.pdf

- Elhabashy, S., factors affecting validity of arterial blood gases results, Germany, LAMBERT Academic Publishing, 2013.

- Elhabashy, S., Formulate Consequential Student Learning Outcomes. , , USA, Johns Hopkins University Press, 2017.

- Elhabashy, S., R. Elkodoos, S. Gebril, B. Osman, M. Elsawy, S. Hamad, N. Kasem, M. Mostafa, A. Mahrous, and M. AbouZead, "A Review of Medical Devices-Related Pressure Ulcers", International Journal of Biology, Pharmacy and Allied Sciences , vol. 7, issue 6, pp. 1133-1146, 2018. httpsdoi.org10.31032ijbpas20187.6.4473.pdf

- Estryn-Behar M (1997) Stress et souffrance des soignants à l'hôpital : Reconnaissance, analyse et prévention, Paris, Éditions Estem.

- F. Sharif and S. Masoumi, "A qualitative study of nursing nurse experiences of clinical practice," BMC Nursing, vol. 4, no. 1, article 6, 2005.

- Fondation canadienne de recherche sur les services de santé (FCRSS) (2006) Les maux qui affligent nos infirmières Examen des principaux facteurs qui portent une incidence sur les ressources humaines infirmières au Canada.

- Gaber Sh., Elhabashy S. "Applied Advocacy for Healthcare Professionals", Advances in Bioresearch, 4(6),2018 pp. 167–175.

- Gaber, S., Global Citizenship in Nursing, USA, Stanford University Press, 2017.

- Gaber, S., F. abed, and I. Said, "Effect of a Developed Evidence-Based Discharge Protocol on Cancer Colon Patients Satisfaction ", Impact Journals, vol. 4, issue 9, pp. 167-176, 2016.

- Gaber, S., Nursing as a Profession and Patient Leading, Guidance, & Support, USA, Yale University Press, 2016.

- Gaber, S., Disaster management at Health Care Settings Comprehensive assessment and effective mitigation, , U.S.A, LULU. Press, 2015. final_shreen_book-1.pdf

- Gaber, S., and N. Fekry, ARE Emergency Nurses Well Prepared for Disaster Management? , Germany, LAP LAMBERT Academic Publisher , 2015. 978-3-659-67138-8_coverpreview-3.pdf

- Gibson V (1994) Does nurse turnover mean nurse wastage in intensive care units? Intensive and Critical Care Nursing 10: 32-40.

- Gilbert MA (1995) Psychodynamique du travail et syndicalisme dans Plaisir et souffrance : Dualité de la santé mentale au travail, Actes du colloque Les aspects sociaux et psychologiques de l'organisation du travail. May 1994. Montreal, Acfas, 66-72.

- Grout JW (1980) Occupational stress of intensive care nurses and air traffic controllers: review of related studies. Journal of Nurse Education 19: 8-14.

- Gurses AP, Carayon P (2007) Performance Obstacles of Intensive Care Nurses. Nursing Research 56: 185-194.

- H. Graneheim and B. Lundman, "Qualitative content analysis in nursing research: concepts, procedures and measures to achieve trustworthiness," Nurse Education Today, vol. 24, no. 2, pp. 105–112, 2004.

- Hatcher J, Bleich MR, Connoly C, Davis K, O'Neill Hewlett P, Stokley Hill K (2006) Wisdom at work: the importance of the older and experienced nurse in the workplace. Princeton, NJ: Robert Wood Johnson Foundation.

- Hay D, Oken D (1972) The psychological stresses of ICU nursing. Psychological Medicine, 23: 109-118.

- Hays M, All AC, Mannahan C, Wallace D (2006) Reported Stressors and Ways of Coping Utilized by

Intensive Care Unit Nurses. Dimensions of Critical Care Nursing 25: 185-193.

- Henderson, S. (1995) Clinical teaching involves more than evaluating nurses. In 4 Annual Teaching Learning Forum.

- Hinchliff S. (1986) The process of clinical nursing. In Hinchliff, S. Teaching clinical nursing (pp. 5-30). Edinburgh, UK: Churchill Livingstone.

- http://www.hrsa.gov/advisorycommittees/bhpradvisory/nacnep/Reports/eighthreport.pdf

- Huckabay LMD, Jagla B (1979) Nurses' stress factors in the intensive care unit. Journal of Nursing Administration 9: 21-26.

- Ismail, M. S., A. Y. Zakaria, Taema, and S. Elhabashy, "ENDOTRACHEAL TUBE PRESSURE INJURY: NURSING PREVENTIVE MEASURES", IMPACT : International Journal of Research in Applied, Natural and Social Sciences (IMPACT : IJRANSS) , vol. 5, issue 10, pp. 9-16, 2017. 3abstract.docx

- J. Baltimore, "The hospital clinical preceptor: essential preparation for success," Journal of Continuing Education in Nursing, vol. 35, no. 3, pp. 133–140, 2004.

- J. Roberts, "Development of a positive professional identity: liberating oneself from the oppressor within," Advances in Nursing Science, vol. 22, no. 4, pp. 71–82, 2000.

- J. Shen and J. Spouse, "Learning to nurse in China—structural factors influencing professional development in practice settings: a phenomenological study," Nurse Education in Practice, vol. 7, no. 5, pp. 323–331, 2007.

- Johnson M (2007) Commentary on Gillespie M &McFetridge B (2006) Nurse education: the role of the nurse teacher. Journal of Clinical Nursing 15, 639-644. J ClinNurs 16: 2178-2179.

- Jokar F. F., Haghani. (2005), nursing Clinical education, challenges: review article. *Iranian Journal of Medical Education*, 10 (5), 1153-1160.

- Jonsén, H.-L. Melender, and Y. Hilli, "Finnish and Swedish nurses' experiences of their first clinical practice placement—a qualitative study," Nurse Education Today, vol. 33, no. 3, pp. 297–302, 2013. View at Publisher • View at Google Scholar • View at Scopus

- Kilminster S, Cottrell D, Grant J, Jolly B (2007) AMEE Guide No. 27: Effective educational and clinical supervision. Med Teach 29: 2-19.

- L. Nahas, "Humour: a phenomenological study within the context of clinical education," Nurse Education Today, vol. 18, no. 8, pp. 663–672, 1998. View at Publisher • View at Google Scholar • View at Scopus

- Lazarus RS (1966) Psychological stress and the coping process, New York, Mc Graw-Hill.

- Le Blanc PM, De Jonge J, De Rijk AE, Schaufeli WB (2001) Well-being of intensive care nurses (WEBIC): a job analytic approach. Journal of Advanced Nursing. 36: 460-470.

- Legault Faucher M (2005) Le travail, la subjectivité, le sujet et l'acteur Attention, ne pas séparer! Prévention au travail, 46-47.

- Li J, & Lambert VA (2008) Job satisfaction among intensive care nurses from the People's Republic of China. Int. Nurs. Rev 55: 34-39.

- Lindberg EB (2007) Increased Job Satisfaction After Small Group Reflection on an Intensive Care Unit. Dimensions of Critical Care Nursing, 26: 163-167.

- Lopez V (2003) Clinical teachers as caring mothers from the perspectives of Jordanian nurses. Int J Nurs Stud 40: 51-60.

- M. Begley and P. White, "Irish nurses' changing self-esteem and fear of negative evaluation during their preregistration programme," Journal of Advanced Nursing, vol. 42, no. 4, pp. 390–401, 2003.

- M. Nabolsi, A. Zumot, L. Wardam, and F. Abu-Moghli, "The experience of Jordanian nurses in their clinical practice," Procedia—Social and Behavioral Sciences, vol. 46, pp. 5849–5857, 2012. View at Publisher • View at Google Scholar

- Maloney J, Bartz C (1983) Stress-tolerant people: intensive care nurses compared with non-intensive care nurses. Heart Lung 12: 389-394.

- Maloney JP (1982) Job stress and its consequences on group of intensive care and non-intensive care nurses. Advances in Nursing Science 4: 31-42.

- Mark BA, Hagenmueller AC (1994) Technological

and environmental characteristics of intensive care units: implications for job redesign. Journal of Nursing Administration 24: 65-71.

- Meltzer LS, Huckabay LM (2004) Critical Care Nurses' Perceptions of Futile Care and its Effect on Burnout. American Journal of Critical Care 13: 202-208.

- Miranda EF (1980) Stressful situations and coping mechanisms of intensive care unit nurses at X hospital. Papers 16: 21-28.

- Mishra J, Morrisey M (1990) Trust in employee/employer relationships: A survey of West Michigan managers. Public Personnel Management 19: 443-463.

- Morgan S. (1991) Teaching activities of clinical instructors during the direct client care period: a qualitative investigation. Journal of Advanced Nursing, P.1238-1246.

- Morrison AL, Beckmann U, Durie M, Carless R, Gillies DM (2001) The effects of nursing staff inexperience (NSI) on the occurrence of adverse patient experiences in ICUs. Australian Critical Care, 14: 116-21.

- N. Hanifi, S. Parvizy, and S. Joolaee, "The miracle of communication as a global issue in clinical learning motivation of nurses," Procedia—Social and Behavioral Sciences, vol. 47, pp. 1775–1779, 2012.

- Nasiri, "Nursing educators and nurses attitude about the effective factors on nursing clinical skill learning in Birjand city 1382," Iranian Journal of Medical Education, vol. 10, article 144, 2004.

- Oerman., F. Garvin. (2002), Stresses and challenges for new graduates in hospitals. Nurse Education .Today, 22(3), 225-230.

- Ordre des infirmières et infirmiers du Québec (OIIQ) (2008) Bulletin d'information des conseils des infirmières et infirmiers.

- Oskin SL (1979) Identification of situational stressors and coping methods by intensive care nurses. Heart & Lung 8: 953-960.

- Pauchant T, Mitroff I (1995) Le niveau existentiel : L'individu et ses défenses. In La gestion des crises et des paradoxes, Montreal, Québec Amérique, ch. 7: 87-101.

- Robinson JA, Lewis DJ (1990) Coping with ICU work-related stressors: a study. Critical Care Nurse 5: 80-88.

- Ruggiero JS (2005) Health, Work Variables, and Satisfaction among Nurses. Journal of Nursing Administration 35: 254-263.

- S. Joolaee, S. R. Jafarian Amiri, M. A. Farahani, and S. varaei, "Iranian nurses' preparedness for clinical training: a qualitative study," Nurse Education Today, vol. 35, no. 10, pp. e13–e17, 2015.

- S. Sheu, H.-S. Lin, and S.-L. Hwang, "Perceived stress and physio-psycho-social status of nurses during their initial period of clinical practice: the effect of coping behaviors," International Journal of Nursing Studies, vol. 39, no. 2, pp. 165–175, 2002.

- S. Speziale, H. J. Streubert, and D. R. Carpenter, Qualitative Research in Nursing: Advancing the Humanistic Imperative, Lippincott Williams & Wilkins, Baltimore, Md, USA, 2011.

- Sawatzky JV (1996) Stress in critical care nurses: Actual and perceived. Heart & Lung 25: 409-417.

- Shirey MR, Fisher ML (2008) Leadership Agenda for Change toward Healthy Work Environments in Acute and Critical Care. Critical Care Nurse 28: 66-79.

- Stone PW, Gershon RR (2006) Nurse work environments and occupational safety in intensice care units. Policy, Politics, & Nursing Practice 7: 240-247.

- Stone PW, Larson EL, Mooney-Kane C, Smolowitz J, Lin SX, et al. (2006) Organizational Climate and Intensive Care Unit Nurses' Intention to Leave. Critical Care Medicine 34: 1907-1912.

- Stone PW, Mooney-Kane C, Larson EL, Pastor DK, Zwanziger J, et al. (2007) Nurse working Conditions, Organizational Climate, and Intent to Leave in ICU: An Instrumental Variable Approach. Health Serv Res 42: 1085-104.

- Trudel L (2000) S'engager dans une enquête en psychodynamique du travail : réflexions méthodologiques. In M-C. Carpentier-Roy & M. Vézina (Eds.), Le travail et ses malentendus. Quebec: Les presses de l'Université Laval, 42-52.
- Tummers GER, Van Merode GG, Landeweerd JA (2002) The diversity of work differences, similarities, and relationships concerning characteristics of the organization 39: 841-855.
- Vézina M (1999a) Organisation du travail et santé mentale : état des connaissances et perspective d'intervention. Revue de médecine au travail 26: 14-24.
- Vézina M (1999b) Stress et psychodynamique du travail: de nouvelles convergences, Travailler, Revue internationale de psychopathologie et de psychodynamique du travail 1: 201-218.
- Vézina M (2000) Les fondements théoriques de la psychodynamique du travail.
- Vézina M, Carpentier-Roy MC (2000) Discussion générale et conclusion. In M.C. Carpentier Roy & M. Vézina (Eds.), Le travail et ses malentendus. Saint- Nicolas, Quebec, Les Presses de l'Université Laval, 147-155.
- Vézina M, Malenfant R (1995) Dualité de la santé mentale au travail. In Plaisir et souffrance : Dualité de la santé mentale au travail, Actes du Colloque Les

aspects sociaux et psychologiques de l'organisation
du travail, May 1994, Montreal, ACFAS, 5-9.

- Weisman, C. (2007). Overview of nursing practice models.
  Retrieved   October   15,   2008,   from,
  http://www.nursinglink.com/training/articles/967-overview-of-
  nursing-practice-models

- White D, Tonkin J (1991) Registered nurse stress in
  intensive care units: an Australian perspective.
  Intensive Care Nursing 7: 45-52.

- Whitman N., Graham B., Gleit C. & Boyd M. (1992)
  Teaching in nursing practice (2nded.). East Norwalk:
  Appleton & Lange.

- Youngblood. N., Janice. M.B., (2001). Developing Critical
  Thinking with Active Learning Strategies. *Nurse Educator*, 26
  (1),

- Z. Mohebbi, M. Rambod, F. Hashemi, H. Mohammadi, G.
  Setoudeh, and D. S. Najafi, "View point of the nurses on
  challenges in clinical training, Shiraz, Iran," Hormozgan
  Medical Journal, vol. 16, no. 5, pp. 415–421, 2012.

- Zakaria, A. Y., Taema, K. M., Ismael, M. S., Elhabashy, S.
  (2018). Impact of a suggested nursing protocol on the
  occurrence of medical Device-related pressure ulcers in
  Critically Ill Patients. Central European Journal of Nursing and
  Midwifery,   9(4),   924-931.   doi:
  10.15452/CEJNM.2018.09.0025